St. George Hotel Complex
16-Alarm Fire
Brooklyn, New York

Investigated by: Scott M. Howell

This is Report 108 of the Major Fires Investigation Project conducted by Varley-Campbell and Associates, Inc. under contract EMW-94-C-4423 to the United States Fire Administration, Federal Emergency Management Agency.

Department of Homeland Security
United States Fire Administration
National Fire Data Center

U.S. Fire Administration Fire Investigations Program

The U.S. Fire Administration develops reports on selected major fires throughout the country. The fires usually involve multiple deaths or a large loss of property. But the primary criterion for deciding to do a report is whether it will result in significant "lessons learned." In some cases these lessons bring to light new knowledge about fire--the effect of building construction or contents, human behavior in fire, etc. In other cases, the lessons are not new but are serious enough to highlight once again, with yet another fire tragedy report. In some cases, special reports are developed to discuss events, drills, or new technologies which are of interest to the fire service.

The reports are sent to fire magazines and are distributed at National and Regional fire meetings. The International Association of Fire Chiefs assists the USFA in disseminating the findings throughout the fire service. On a continuing basis the reports are available on request from the USFA; announcements of their availability are published widely in fire journals and newsletters.

This body of work provides detailed information on the nature of the fire problem for policymakers who must decide on allocations of resources between fire and other pressing problems, and within the fire service to improve codes and code enforcement, training, public fire education, building technology, and other related areas.

The Fire Administration, which has no regulatory authority, sends an experienced fire investigator into a community after a major incident only after having conferred with the local fire authorities to insure that the assistance and presence of the USFA would be supportive and would in no way interfere with any review of the incident they are themselves conducting. The intent is not to arrive during the event or even immediately after, but rather after the dust settles, so that a complete and objective review of all the important aspects of the incident can be made. Local authorities review the USFA's report while it is in draft. The USFA investigator or team is available to local authorities should they wish to request technical assistance for their own investigation.

This report and its recommendations were developed by USFA staff and by Varley-Campbell & Associates, Inc., Miami and Chicago, its staff and consultants, who are under contract to assist the Fire Administration in carrying out the Fire Reports Program.

The U.S. Fire Administration greatly appreciates the cooperation received from officials of the City of New York Fire Department, most particularly Commissioner Howard Safir and Assistant to the Commissioner Tom Fitzpatrick.

For additional copies of this report write to the U.S. Fire Administration, 16825 South Seton Avenue, Emmitsburg, Maryland 21727. The report is available on the Administration's Web site at http://www.usfa.dhs.gov/

U.S. Fire Administration
Mission Statement

As an entity of the Department of Homeland Security, the mission of the USFA is to reduce life and economic losses due to fire and related emergencies, through leadership, advocacy, coordination, and support. We serve the Nation independently, in coordination with other Federal agencies, and in partnership with fire protection and emergency service communities. With a commitment to excellence, we provide public education, training, technology, and data initiatives.

 FEMA

TABLE OF CONTENTS

OVERVIEW 1

KEY ISSUES 2

THE FIRE AREA 2

THE FIRE 3

GREATER ALARMS 5

ADDITIONAL ALARMS 6

TOWER BUILDING A 6

EXPOSURES 6

CONTROL 7

SPECIALIZED UNITS 7

    Communications 7

    Water Supply 8

    Other Specialized Units 8

LESSONS LEARNED AND REINFORCED 8

APPENDIX A: Diagrams 11

APPENDIX B: Photographs 22

St. George Hotel Complex 16-Alarm Fire
Brooklyn, New York
August 26, 1995

Investigated by: Scott M. Howell

Edited by: Sheila-Faith Barry

Local Contacts: (Former) Commissioner Howard Safir
                  Tom Fitzpatrick, Assistant to the Commissioner
                  Assistant Chief Edward Cunningham
                  Deputy Assistant Chief Thomas Lally
                  Deputy Chief David Corcoran
                  City of New York Fire Department
                  250 Livingston Street
                  New York, New York 11201-5884
                  (718) 694-2056

OVERVIEW

More than 700 firefighters operating over 100 pieces of firefighting apparatus were needed to control a fire that involved several large interconnected buildings in a crowded neighborhood in Brooklyn, New York on August 26, 1995. The complex of buildings was known as the St. George Hotel. It was the largest fire in New York City in more than twenty years, and one of the largest in the city's history.

The fire began in a vacant nine-story building and spread to several adjacent exposures, including an occupied 31-story apartment building. The fire was well developed on the ninth floor by the time it was located by the first arriving companies due to a delay in reporting the fire and the initial dispatch to a different building. An interior attack was attempted, however, the standpipe in the fire building was not functional and the firefighters were forced to abandon their initial attack as the fire spread through the original fire building and into the exposures.

The fire presented severe life safety and property exposure challenges on all four sides, in addition to generating a storm of fire brands that were carried by the wind and threatened an even larger area.

KEY ISSUES

Issues	Comments
Delay in Reporting	A delay in discovery and reporting of the fire allowed the fire to spread to a larger area before firefighters arrived.
Protection of Vacant Buildings	The failure of the standpipe in the fire building compromised the initial attack and endangered personnel. Rigorous inspections and enforcement of codes requiring fire protection systems to be maintained and vacant buildings to be secured against vandalism could have prevented this situation.
Need to call for Resources Early	Because of massive exposure problems this fire required a tremendous commitment of resources. The Incident Commander's decision to call for additional resources in anticipation of the need was a major factor in the successful outcome of this incident.
Fire Risk of Vacant Buildings	Vacant buildings and complexes in crowded neighborhoods present a large fire risk.
Incident Command System	The Fire Department of New York City has a wealth of experience in conducting major operations and has developed standardized procedures for implementing components of the Incident Command System.

THE FIRE AREA

The St. George Hotel was the grandest of New York City's hotels when it opened in 1926. Located in the Brooklyn Heights area of the city, it had a total of 2,632 guestrooms in a complex of eight buildings that occupied an entire city block. (Figure 1). Some of the original buildings dated back to 1855. All the buildings were interconnected at the basement level, and there were numerous connections on various floor levels among the different buildings.

The complex was bordered by Pineapple Street to the north, Clark Street to the south, Hicks Street to the east and Henry Street on the west. Many streets in the Brooklyn Heights area are only 40 feet wide. Parking on the streets in front of the fire building further congested the area. Figure 1 shows the block and the relationships of the buildings to each other.

Over the years, several of the buildings in the complex were vacated, while others were converted to residential apartment use. Only the original St. George Hotel building on the east end of the complex operated as a hotel at the time of the fire.

The building of fire origin, known as the Clark Building, was located at 51 Clark Street. It was nine stories high, measured 75 feet wide by 90 feet deep, and had been vacant for at least seven years.

The ten story Grill Building was immediately west of the Clark Building on the corner of Clark and Hicks. Although the Grill Building was only 25 feet deep, it had 100 feet of frontage on Clark Street. At one time the Grill Building and the Clark Building had been connected at each floor level with overhead rolling fire doors at the openings. After the Clark Building was vacated the fire doors were closed and the openings were bricked over on the Grill Building side.

The 31 story Tower Building ran nearly the length of the block on Hicks Street, between Clark and Pineapple Streets behind the Grill and Clark Buildings. The Tower Building had 175 feet of frontage on Hicks Street and 150 feet on Pineapple Street. The main entrance was at 111 Hicks Street. The Tower Building was built in 1929 of protected steel construction and was connected to the Grill Building at the basement level. The portion of the Tower Building that fronts on Pineapple Street was built as a free standing structure known as the St. George Apartments. In recent years the original Tower Building, the old St. George Apartments and the Grill Building had all been combined into one residential building known as the Tower Building.

East of the Tower Building were the six story Crosshall and the eight story Pineapple Buildings. The Crosswalk Building had a fifty-foot frontage on Pineapple Street, while the Pineapple Building at 60 Pineapple Street is 115 feet wide and, like Crosshall, is ninety feet deep.

The building that was being operated as the St. George Hotel occupied the entire east end of the block. The section that faces Pineapple is eight stories tall, while the portion that faced Clark Street was 12 stories tall. There were entrances on Henry Street, which had more than 200 feet of frontage, and on Clark Street, which had 125 feet of frontage. The New York City Transit Authority operates a subway station under the St. George Hotel with direct access from the lobby to the station.

The Weller Building, west of the St. George Hotel, was four stories in height with a 50 by 85 footprint. Between the Weller and the Clark Buildings was the 12 story Marquee Building, which had been vacant for several years. The Marquee Building was separated form the Weller Building by fire doors. The 50 by 85 foot Marquee Building was open to the Clark Building on all floor levels. A common one story covered courtyard occupied the remaining open area between the buildings at the west end of the block.

All the buildings on the block, except the 31 story portions of the Tower Building, were of ordinary construction. The Clark and Marquee Buildings were both vacant at the time of the fire while all of the other buildings were being sued as residential occupancies. The St. George Hotel had some permanent residents and was also being used by various city agencies to provide temporary housing for AIDS patients. Standpipe systems and partial sprinkler systems were present and functional in the occupied building.

The vacant Clark Building had been legally and illegally scavenged. Large sections of flooring were missing and the standpipe system was partially dismantled. Vagrants were known to frequent the building despite efforts to keep the building sealed as required by New York City codes.

THE FIRE

The weather was warm in New York City in the early morning hours of August 26, 1995 and many of the occupants of the St. George Hotel and the surrounding buildings had their windows open for ventilation. Sometime close to 3:00 a.m. several occupants called down to the hotel desk to report a smell of smoke. The phones in the rooms did not have the capability to call outside the hotel itself. The hotel staff noted that the alarm system for the building was not indicating a problem so they did not call the fire department.

At 3:31 a.m. the FDNY Brooklyn Communications Center received a call reporting smoke in the area of the St. George Hotel on Henry Street. A first alarm response consisting of three engines, two trucks and a battalion chief was dispatched and were on the scene by 3:36 a.m. Upon entering they were told there was no problem in the hotel and that the fire alarm control panel showed no activation. The battalion chief (BC31) asked the dispatcher to call back the person reporting the smoke to get a better location. The dispatcher's call was answered by an answering machine, so no additional information was available.

The units on the scene followed their Standard Operating Procedures, which included sending the "roofman" of the first due-in truck company to the roof. Once there he was able to see the glow of fire, which was several buildings down the block. At about the same time a civilian reported light smoke visible above the buildings around the corner, on Clark Street, to the crews standing by. At 3:44 a.m. the battalion chief, BC-31, transmitted a radio report to confirm the presence of a fire. An additional Battalion Chief (BC-35) a fourth engine company and a squad company were dispatched.

The crews on the scene repositioned their apparatus on Clark Street and investigated further, but the extent of the fire was not obvious from street level because of the narrow streets and the fire's location in the upper floors at the rear of the ten-story building. Entry was made to the ground floor of the building through a boarded-up doorway. The crews were familiar with the general building layout fro past incidents and pre-fire plans but had trouble gaining access to the stairway because of trash piled up in front of the stairway opening.

At 3:34 a.m. BC-31 transmitted an "all hands" signal with prompted dispatchers to assign a Deputy Chief, (DC-6), a rescue squad and a third truck company. A second alarm and a special call for an additional tower ladder were transmitted about five minutes later by BC-31 as the fire was growing in intensity and spreading rapidly. The Field Communications Unit and Maxi-Water system responded automatically on a second alarm, along with four engine companies and one truck company. The Citywide Tour Commander also responded on the second alarm.

The two first alarm truck companies performed a primary search of the fire building. After making their way up the stairs to the eighth floor, the engine companies connected 2-1/2-inch hand lines to the standpipe and the order was given to charge the system. However, the standpipe had apparently been scavenged and sections of the pipe were missing. The 2-1/2-inch hand lines were then lowered out the windows, hooked up to an engine in front of the building and charged. At this point, the command post was in the street in front of the building. BD-31 was Incident Commander. (Figure 2).

The interior fire attack had to be terminated after a very short time when it was determined that the fire was spreading down to the lower floors via the elevator shaft and other openings in the floors. BC-35 gave the order to remove the attack crews from the interior of the building. The fire was spreading so rapidly that the standpipe packs were abandoned.

At this time Tower Ladder 119 was setting up in front of the building in anticipation of an exterior attack on the fire. TL-119 was ordered not to begin operating until all the members who had been working inside the Clark Building were outside. As the last of the firefighters were leaving the fire building, the fire dropped all the way down to the street level and flames were visible on all ten floors. There was heavy involvement of the eighth and ninth floors. The firefighters in TL-119's basket were experiencing extreme radiant heat as they waited for the order to open up their master stream.

At 4:06 the first arriving Deputy Chief, DC-6, ordered a third alarm and assumed incident command (Figures 3 and 4) as the fire was beginning to spread to exposure 4, the Marquee Building. The third alarm brought in four engines, 2 trucks, the Air Mask Maintenance Unit and a Medical Command Unit.

The fire quickly began to produce a huge volume of fire brands. Residents used garden hoses to protect their roofs as the brands landed on nearby structures. Brands floated through open windows into the taller adjacent buildings and ignited fires in apartments. The fire department hoselines on Clark Street had to be washed down for protection.

The Incident Commander assigned BD-35 to Exposure 4 to stop the fire from spreading through the Marquee Building to the attached buildings on the east side. A primary search was performed in the Marquee Building, but due to its poor structural condition crews were instructed to attack that portion of the fire from the next building to the east, the Weller Building (exposure 4A). In essence the vacant Marquee Building was used as a buffer to protect the buildings to the east.

As the additional companies from the "all hands" signal arrived they were assigned to the building with the greatest apparent life hazard, the Tower Building. Some of the apartment windows in this building were less than thirty feet from the fully involved upper floors of the Clark Building. The standpipes in the Grill and the Tower Buildings were charged while companies began the task of searching and evaluating the residents of the 31-story building. Many of the residents had been awakened by the noise of the incoming apparatus or the light of the flames and were self-evacuating.

An additional Deputy Chief was requested to take charge of Exposure 3, the areas to the rear of the fire building on Pineapple Street. Several additional Battalion Chiefs were special called, as the need for more supervisory personnel became critical.

The Citywide Tour Commander FC-1 requested a fourth alarm at 4:11 a.m. while he was still enroute. From the Brooklyn Bridge he had an excellent vantage point to size up the volume of fire and the obvious exposure challenges. By this time the interior portions of the upper floors of the Clark Building were beginning to collapse, adding to the storm of fire brands. (Figures 5 and 6).

When FC-1 arrived on the scene at 4:18 a.m., he immediately called for a fifth alarm. The interior of the original fire building was well involved and the fire had spread to the upper floors of Exposure 4. A collapse zone was quickly established in front of the Clark and Marquee Buildings. The two tower ladders that had been set up in front of the buildings and the command post were repositioned out of the collapse zone. Battalion Chiefs were assigned to secure the collapse zone. (Figures 7 and 8).

Directly across the street from the fire building at 52 Clark Street was Exposure 1, another highrise building in which many older and handicapped people lived. The only exit from this building was onto Clark Street directly across the street from the fire building. Evacuation would require the 450 residents to walk within 40 feet of the front wall of the ten-story fire building. With the risk of collapse and the intense radiant heat the decision was made to keep these residents in their building, but to move them to the rear apartments as far as possible from the danger and protect them in place. This was accomplished with the help of the Police Department.

By this time the Incident Command System was well established. The Command Post was positioned on Clark Street east of the fire building. FC-1 was the Incident Commander, the first-in Deputy Chief DC-6 was the Operations Chief, and Deputy Chief DC-1 was in position on Pineapple Street to cover Exposure 3. Battalion Chiefs were assigned to both ends of the collapse zone to keep the danger area clear and to direct the tactical operations in their sector. Exposure 4 was a vacant building that was becoming involved in fire but was acting as a buffer for the occupied structures to the west. Exposures 1, 2 and 3, which were all occupied structures and significantly taller than the fire building, were all critical exposures.

GREATER ALARMS

It was obvious that the 31 story Tower Building that faced Pineapple Street was in immediate danger. The Incident Commander directed the fourth and fifth alarm companies to report to the Chief on exposure 3 to assist in evacuating and protecting the Tower Building.

A Battalion Chief was assigned to conduct a survey of the entire block. It appeared from the street that the Grill, Tower and the St. George Apartments were separate buildings.[1] However, the survey determined that the interior had been inter-connected to create one large structure.

---

[1] The landmark status of the area mandated that the exterior of the buildings remain basically the same as they had been before they were consolidated.

When the Incident Commander received this information he informed DC-1 of the building configuration and directed him to take charge of both Exposures 2 and 3 (Figure 9). The fireground was now divided into the Clark Street Branch, which included Exposures 1 and 2 and the fire building, and the Pineapple Street Branch, which included Exposures 2 and 3. Two Deputy Chiefs were assigned to direct operations in the two branches, DC-6 on the Clark side and DC-1 on the Pineapple side. The Incident Commander directed DC-1 to initiate a separate incident and to communicate directly with Brooklyn Communications to request resources.

ADDITIONAL ALARMS

A secondary Command Post was established at the corner of Hicks and Pineapple and a second alarm assignment was requested by DC-1 to report to the Pineapple Sector, which included the Pineapple and Crosshall Buildings. The equivalent of a third alarm was requested to assist the companies that were already operating in the Tower Building (Figure 10).

Additional alarms for the Clark Street side were transmitted as follows: the sixth at 5:06 a.m., the seventh at 5:09 a.m., the eighth at 5:28 a.m., the ninth at 6:15 a.m., the tenth at 7:17 a.m. and the eleventh at 7:20 a.m. The fire was declared under control at 7:09 a.m. (Figure 11).

TOWER BUILDING A

A primary search of the exposures was a major challenge due to the large number of living units exposed. Although most of the residents of the Tower Building self-evacuated, the concierge was receiving calls from numerous residents requesting help. Other residents called 9-1-1 and their locations were relayed to the command post by Brooklyn communications. The Sector Commander prioritized these requests and assigned companies as they arrived or became available. Fires in several apartments had been ignited by the flying brands or radiant heat. In an effort to reach the areas most in need of intervention, apartment doors were forced open if heat was detected on the corridor side of the door. Entry ultimately had to be forced into 84 apartments in the Tower Building.

As more Chief Officers arrived, the Tower Building was divided into sectors. Several floors and companies were assigned to each Battalion Chief.

Some of the first companies into the Tower Building used the elevators to reach the upper floors. Some of the later arriving companies had to climb 20 to 25 stories to reach their assigned areas. The building's two stairways became congested with hundreds of residents evacuating down while firefighters were advancing up the stairs, carrying their equipment. Smoke in the corridors and stairways added to the confusion. A specially trained highrise crew was sent to retrieve the elevators from upper floors and to assure their proper use.

As more resources arrived and were assigned, a total occupant search was completed and all of the apartment fires were extinguished. Sixteen 2-1/2-inch lines were stretched from the standpipes into apartments to directly attack the fire in the Clark Building. The radiant heat from so large a fire continued to present a major threat and to tax the resources on the scene.

EXPOSURES

The fire brand problem was multiplied as the floors of the Clark Building collapsed. Flaming brands found their way into the open or failed windows of apartments and landed on the rooftops of surrounding buildings. Exposed apartments had to be checked and rechecked to ensure that the contents did not ignite. Two companies and a Battalion Chief were assigned as a "Brand Patrol Group" for the rest of the neighborhood.

Two apartments at 60 Pineapple Street were found in flames and were extinguished. After a search and evacuation of residents was accomplished, additional hose lines were operated from the roof of this building into the rear of the Clark and Marquee Buildings.

Hoselines were also operated from the roof and from all floors of the Weller Building into the Marquee Building on the Exposure 4 side. The fire doors between the two buildings held until the firefighters could evacuate the Weller Building and get their lines in place to attack the fire.

A tower ladder was set up in the street in front of 60 Pineapple and operated over the rooftop into the fire building. Three additional tower ladders operated into the front of the Clark Building to extinguish the fire. The collapse zone was maintained and fortunately the fire did not collapse significantly outside its footprint.

CONTROL

As the danger form the fire decreased, the residents of Exposure 1 at 52 Clark Street were removed to safety. This building suffered only slight water damage from the hoselines that were used to protect it from the radiant heat.

The masonry bearing walls were all that remained of the original fire building. Virtually all combustible materials had been consumed. Residents of many of the surrounding buildings were not allowed back into their apartments until the walls were demolished due to the collapse danger. Fire companies remained on the scene for several days putting out spot fires.

SPECIALIZED UNITS

Communications

The communication requirements for this Incident were a major challenge. The FDNY made extensive use of portable radios or handi-talkies (HT's) to coordinate tactical operations on the fireground. The HT's are distributed as follows:

> 2 HT's for each Deputy Chief and Battalion Chief
>
> 3 or 4 HT's for each Ladder Company
>
> At least 2 HT's for each Engine, Squad or Rescue Company

At most multiple alarm incidents, the command officers use a Command Channel, while companies use a Tactical Channel. Battalion and Deputy Chief's monitor and communicate on the Command Channel while their aides monitor the Tactical Channel.

The Field Communication Unit (Field Com) responded automatically on the second alarm. Battalion Chief 32 was assigned as communications coordinator at 4:25 a.m. with a listing of all units assigned to the incident up to that point in time.

The Communications Coordinator ensures that all of the units already on the scene are on the proper Tactical and Command channels, then monitors the Tactical channel for urgent transmissions.

The sizes of this incident required both of the tactical channels available in this area of the city. One tactical channel was used for the Clark Street Branch and the other for the Pineapple Street Branch. Only one Command Frequency is available in this area of Brooklyn, a second would have been useful if it had been available.

The Communications Coordinator remains in close contact with the Incident Commander and maintains a status board which keeps record of the Branch and Sector assignments, the units operating in each area and the companies that are available in the staging area. The Command Control Chart is maintained at the Command Post in a graphic format for easy reference.

The Computer Assisted Dispatch Operations (CADO) Unit also responds on the second alarm and reports to the Communications Coordinator.

Water Supply

The Maxi-Water System is dispatched on second alarms in this part of Brooklyn. The entire Maxi-Water System consists of six engine companies with 2000 gpm pumps, each of which has a hose wagon loaded with large diameter hose and appliances. Engine 207 is the primary Maxi-Water company and the officer of E-207 acts as the Water Resource Officer at major incidents. The five Satellite Units are assigned to other areas of the city. The normal response of the Maxi-Water System included Engine 207 as the Maxi-Water Unit and two of the Satellite Units. The 2000 gpm pumpers are directed to hydrants that can provide a strong water supply and delivery water through the large diameter hose to manifolds close to the fire.

Due to the location of fire, E-207 was one of the engines on the initial response and was assigned early in the fire incident.

The Water Department was asked to open the grids to assure maximum water availability in the area. Manifolds were positioned on Clark Street, Pineapple Street and Hicks Street to supply both handlines and heavy streams.

Other Specialized Units

One of the two Safety Operating Battalion chiefs responds automatically on second alarms. He is responsible to survey the scene for safety practices and monitor radio traffic for emergency calls. SB-2 was assigned to this incident.

The extra ladder company that is dispatched on an "all hands" fire is assigned as the Firefighter Assist and Search Team (FAST). The FAST reports to provide immediate assistance in the event that a company gets in trouble or cannot account for a firefighter. Due to the magnitude of this incident, several ladder companies were assigned as FAST companies for different use.

LESSONS LEARNED AND REINFORCED

1. **Updated pre-fire plans provide critical information regarding the configuration and condition of buildings within a complex.**

    Pre-fire plans had been developed for this area in the past but may not have been up to date. The Incident Commander did not have the information that three buildings had been consolidated into one. This caused some confusion but did not significantly effect the outcome of the incident. The standpipe in the fire building was out of service. The fire department has a tagging system to indicate that a standpipe or sprinkler system is out of service. This system had been vandalized since the last inspection.

2. **Walking all perimeters of the fire scene can provide information regarding exposures and conditions, and assist in the effective placement of firefighting resources.**

Because of the confusion about the configuration of the buildings in the complex, the Incident Commander sent a Battalion Chief to scout the area to obtain specific information. This put an end to the inconsistent reports the Command Post was receiving and allowed confident placement of resources.

3. **Aggressive resource staging can minimize the loss and assist with occupant location, evacuation and protection, as well as improving firefighter safety.**

The Incident Commander stated that whenever a problem came up he called an additional alarm. The outcome of the incident was much more positive because the Incident Commander readily brought additional personnel to the scene. Due to the size and location of the fire there was great potential for large loss of life and significantly more property damage than actually occurred. The initial Incident Commander had knowledge of what equipment was on the way as well as what equipment was needed. Making special calls for specific equipment, such as extra tower ladders, almost enabled extinguishment of the fire at the second alarm level.

4. **Specialized units contribute to an effective fire attack.**

Many of the component parts of the Incident Command System were pre-assigned in this fire. Comcord, Maxi-Water and the Safety Battalion are examples of these components. These specialized units were trained to assume responsibility for a very focused and defined task. This well-defined area of responsibility was assigned to the specialized unit. The Incident Commander was confident that work expected from a unit was exactly the work the unit expected to perform.

5. **Pre-assigned radio frequencies and an orderly way to switch to them are vital to success.**

One of the biggest problems in a large incident is usually communications. By necessity, there is a sizable volume of radio traffic. Dispatch or main frequencies will quickly be overloaded if provisions are not made early in an incident to organize radio traffic. By promptly establishing two tactical and one command channel in addition to the dispatch frequency, the ComCord maintained control of communications.

6. **Radio discipline, regardless of the number of fireground radio channels available, should be emphasized by officers and companies.**

Operating 100 pieces of apparatus and 700 firefighting personnel, most of which had hand held radios, on two tactical channels and one command channel required discipline and vigilance. The command officers had aides able to monitor tactical radio communication. Companies kept their radio communications to a minimum so emergency traffic and important information could be passed on.

7. **Gain control of the elevators early in the fire.**

Certain vital building systems need special attention to assure safe operation during a fire. Elevators are one of these systems. The elevators were abandoned on the upper floors of the Tower Building early on during the incident. As a result, subsequent companies had to climb the stairs to reach their operating area. When the problem was relayed to a chief, a company was assigned to retrieve the elevator and operate it for the duration of the incident. If the elevators are controlled early, crews can safely use them for transportation and they will be less fatigued when they reach their operating areas.

8. **Routine inspections and up-to-date pre-incident plans can identify compromised fire protection systems.**

A lot of time was spent by the first-in companies advancing to the fire floor believing they would be able to operate from the standpipe. If the inoperable state of the standpipe had been known, an alternative way of supplying water to the fire floor could have been employed, saving valuable time when it was most needed. The disabled standpipe may have been found during a routine inspection. A pre-incident plan must be continuously updated, as building conditions can change. Inspections should be performed on a regular basis so those changes can be reflected on the pre-incident plans.

9. **Using the Department's Incident Command System (ICS) permitted this complex fireground to be managed to a successful conclusion without loss of life or serious injury to civilians or firefighters.**

The first-in companies and battalion chief followed the Department's Standard Operating Procedures. ICS was established from the beginning of the incident and functioned through command changes, command post relocation and separation of the fireground into two branches, each with multiple sectors. The experience and familiarity of the command and company officers with ICS was evident during this incident. The large fire area and congested exposures required a concentrated and coordinated effort that effective ICS supports.

APPENDIX A

Diagrams

Figure 1. Site Plan

Appendix A (continued)

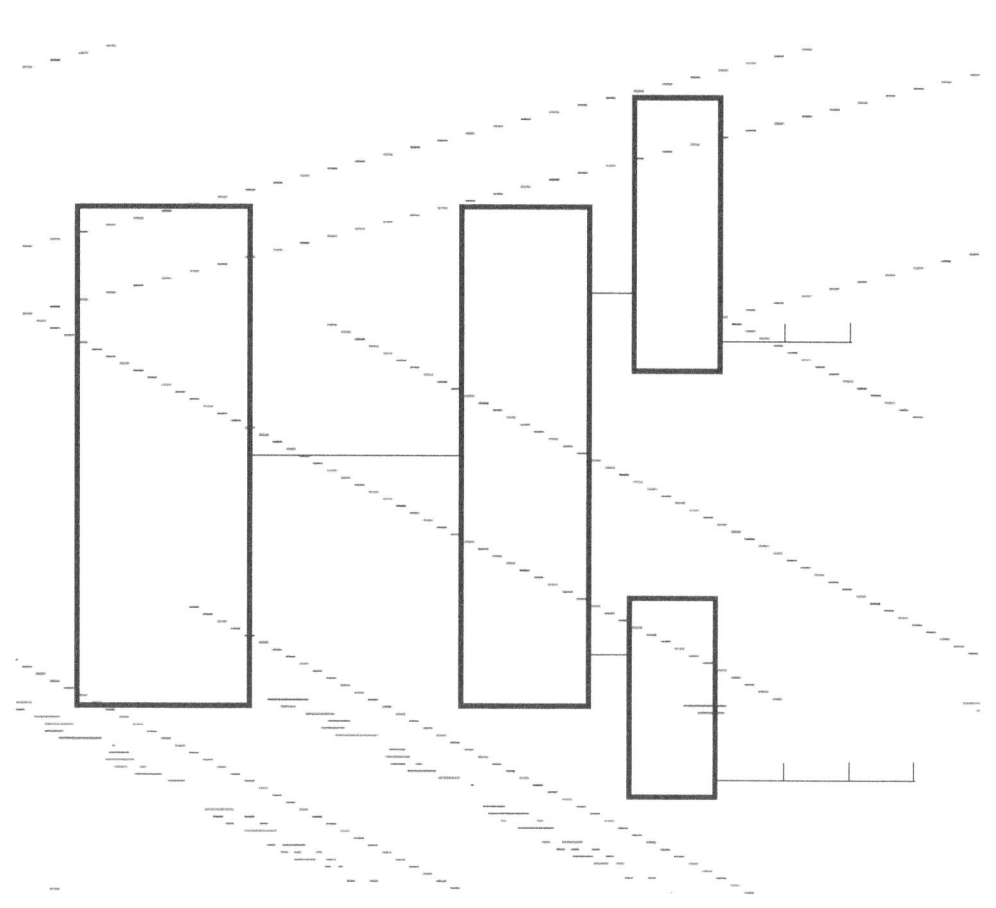

Figure 2. Incident Command Structure at About 3:50 a.m.

Appendix A (continued)

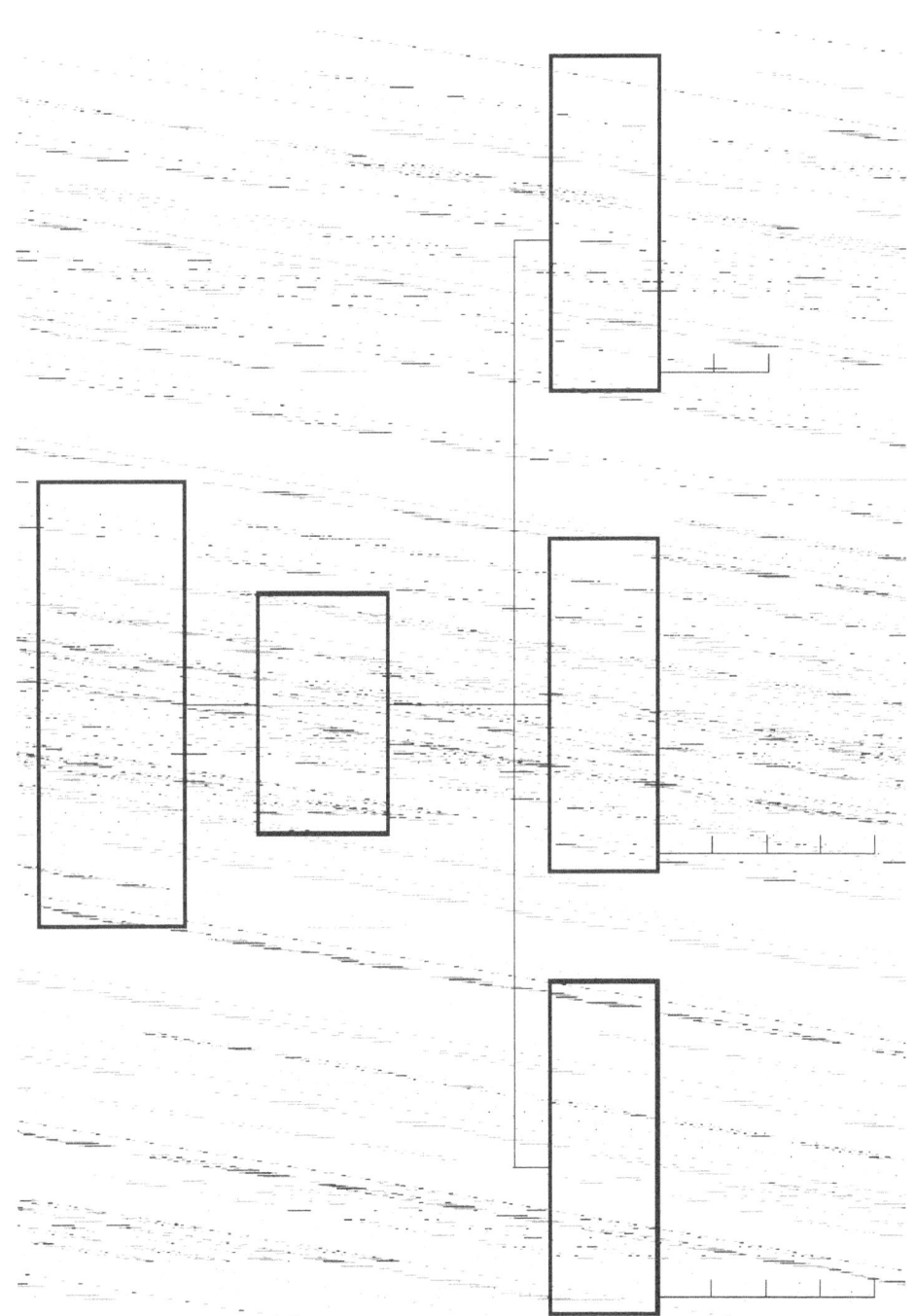

Figure 3. Incident Command Structure at About 4:10 a.m.

Appendix A (continued)

Figure 4. Placement Of "All Hands" Companies at About 4:10 a.m.

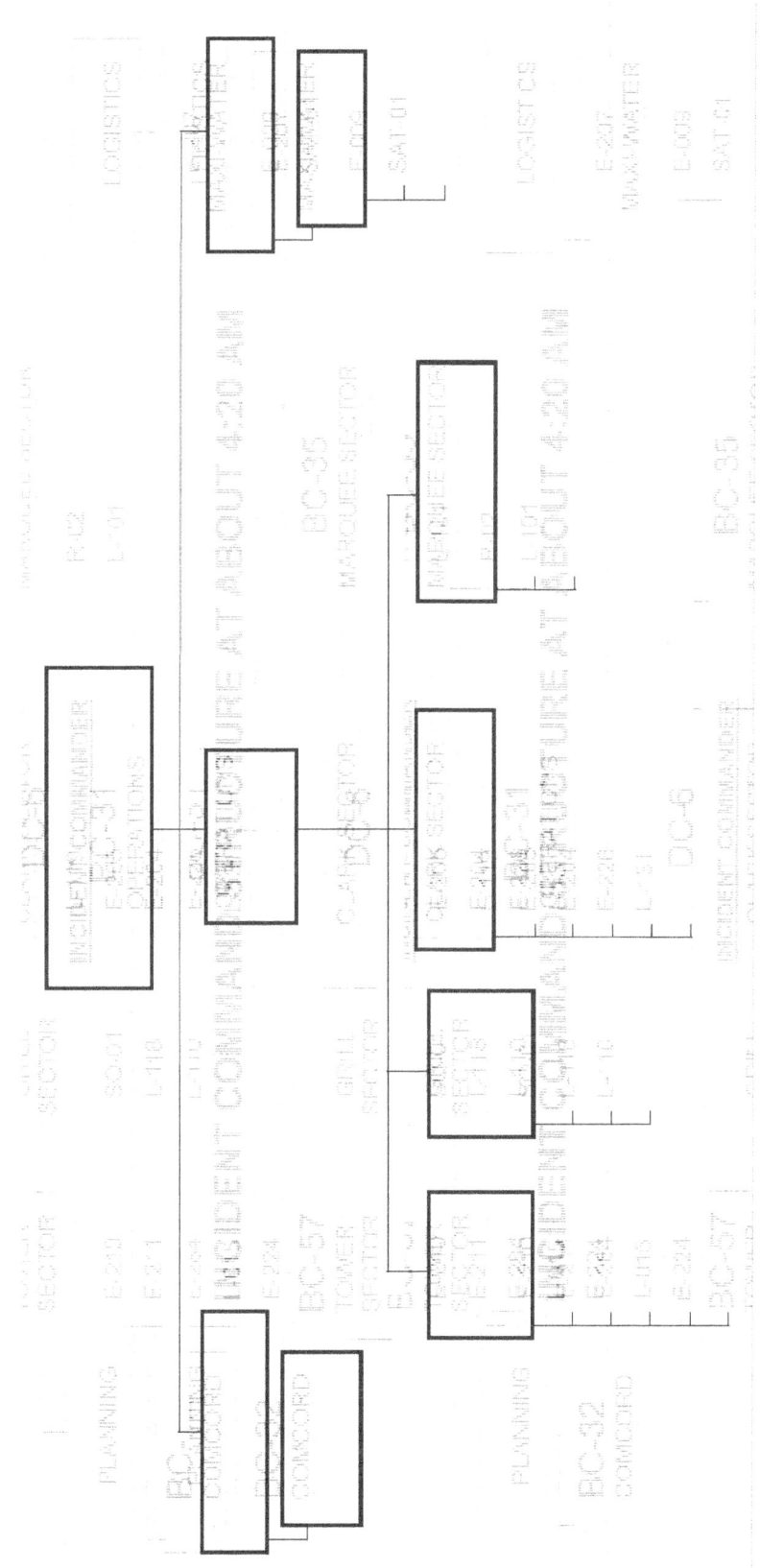

Appendix A (continued)

Figure 5. Incident Command Structure at About 4:20 a.m.

Appendix A (continued)

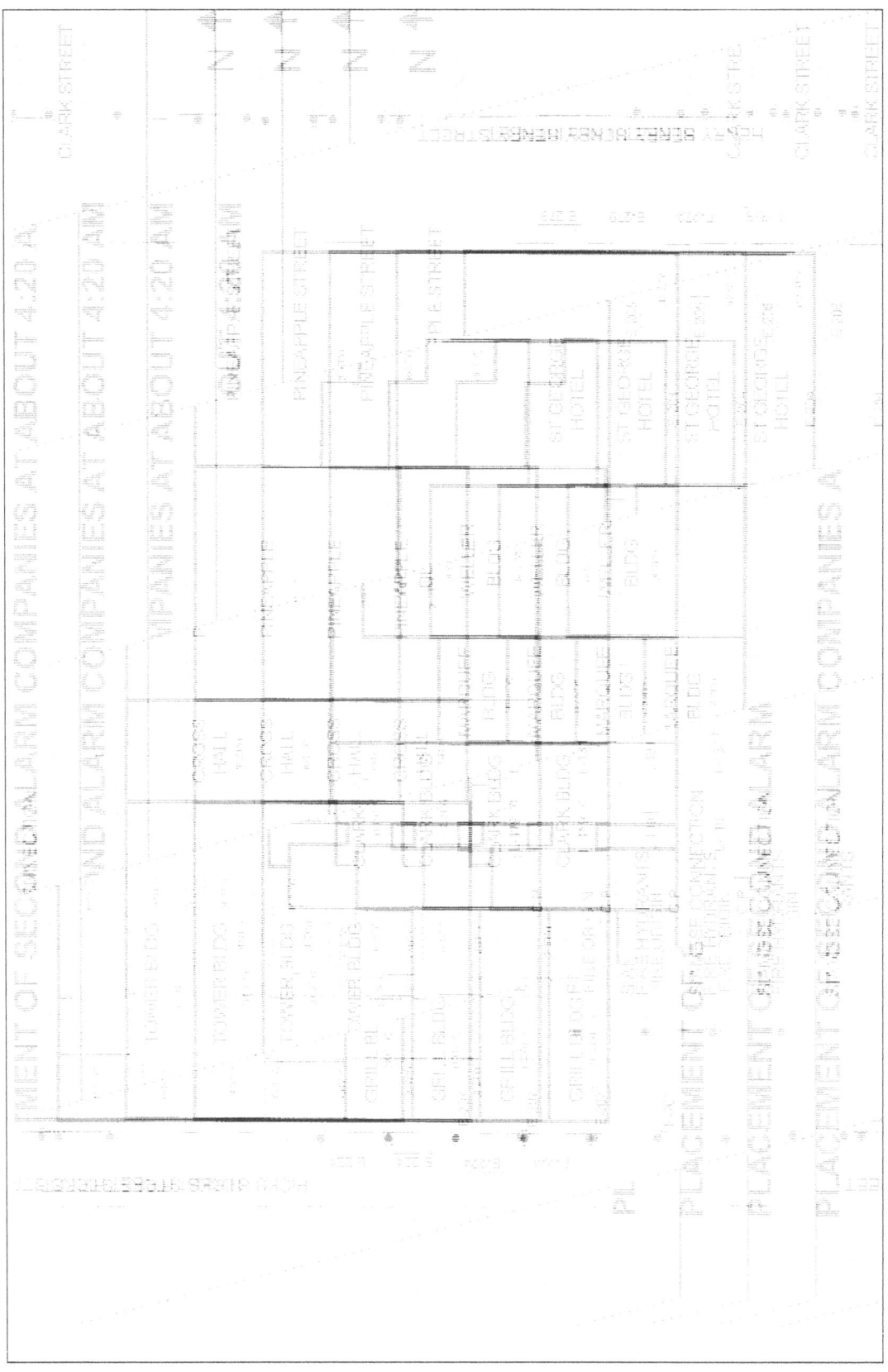

Figure 6. Placement Of Second Alarm Companies at About 4:20 a.m.

Appendix A (continued)

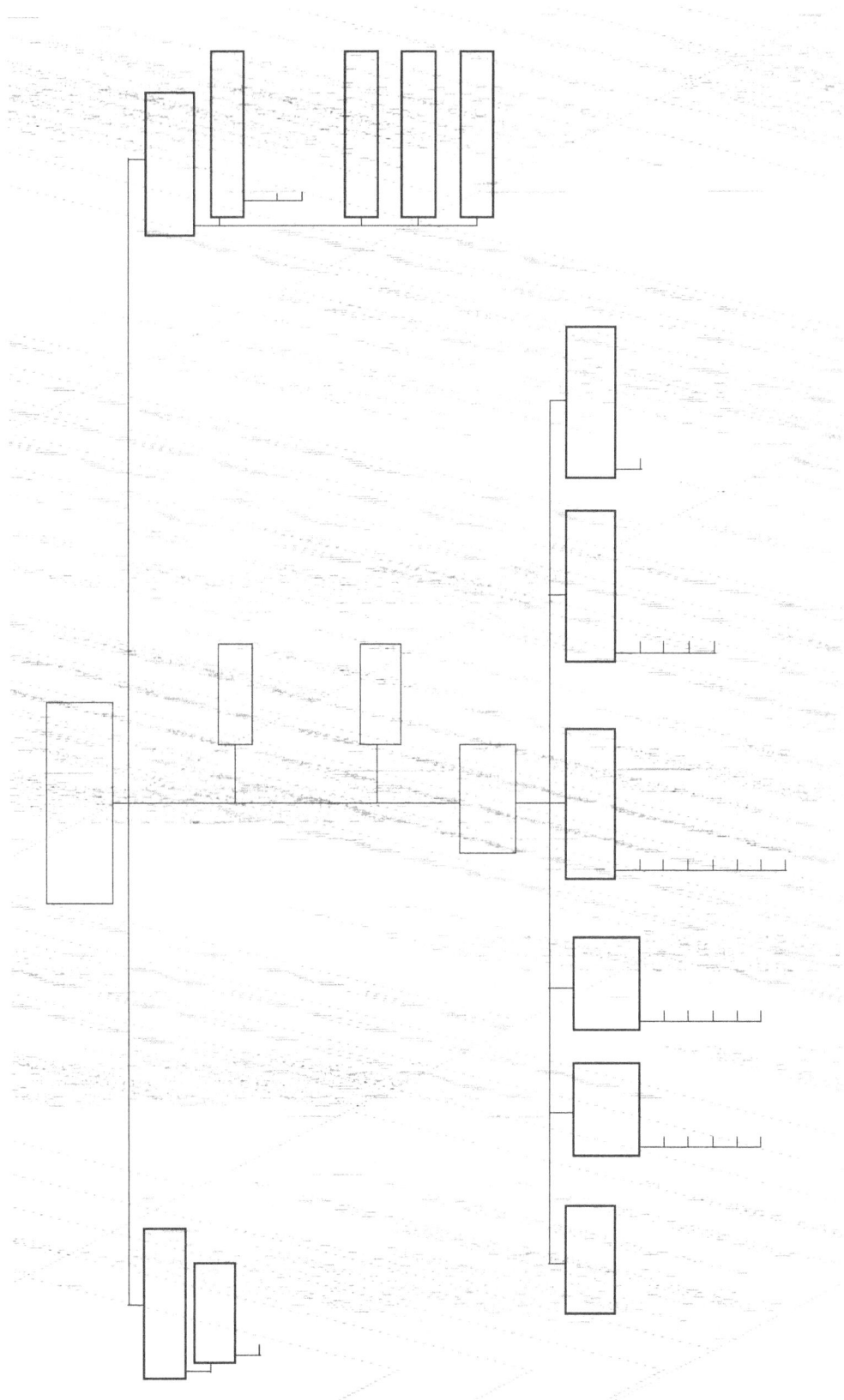

Figure 7. Incident Command Structure at About 4:35 a.m.

Appendix A (continued)

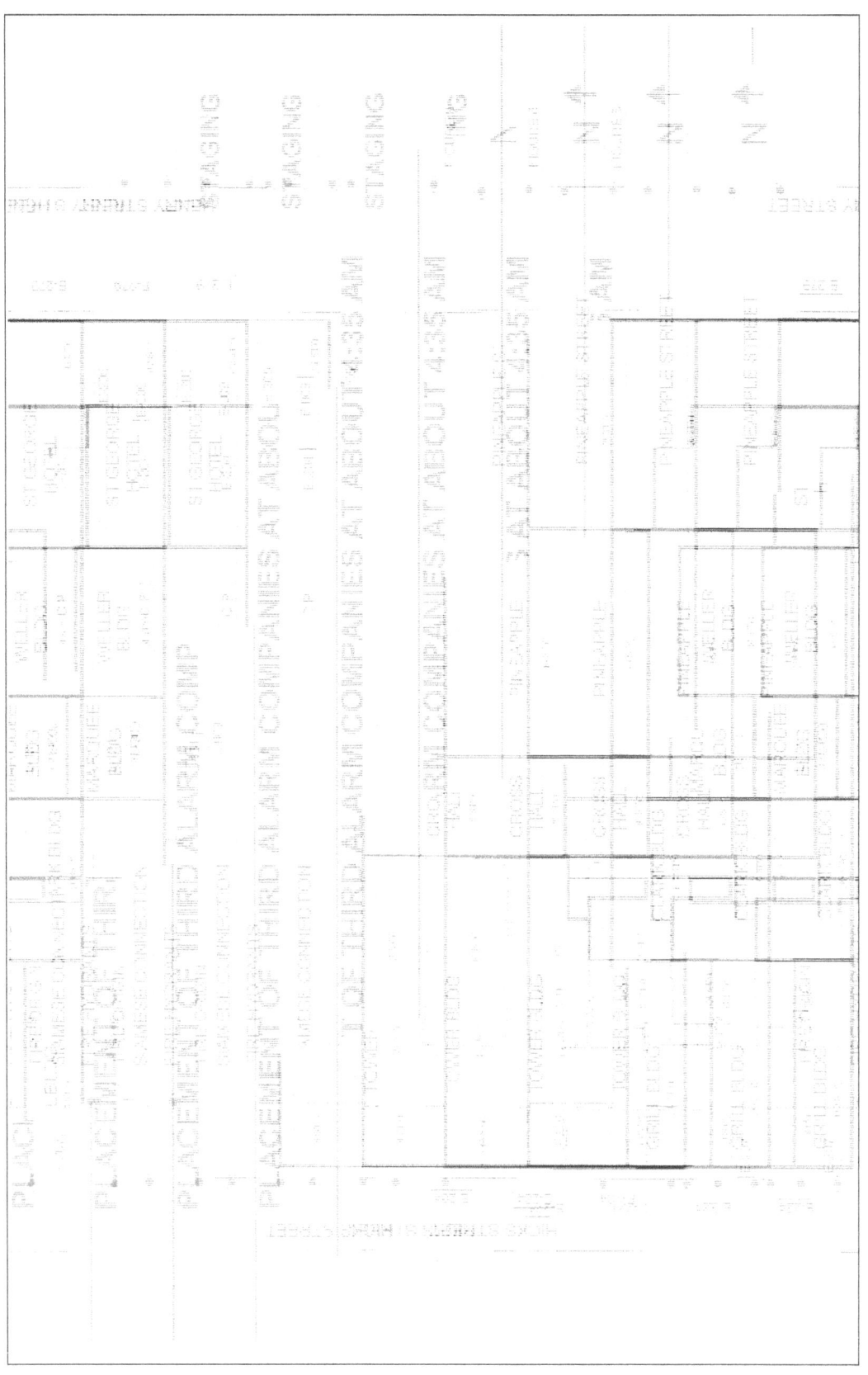

Figure 8. Placement Of Third Alarm Companies at About 4:35 a.m.

Appendix A (continued)

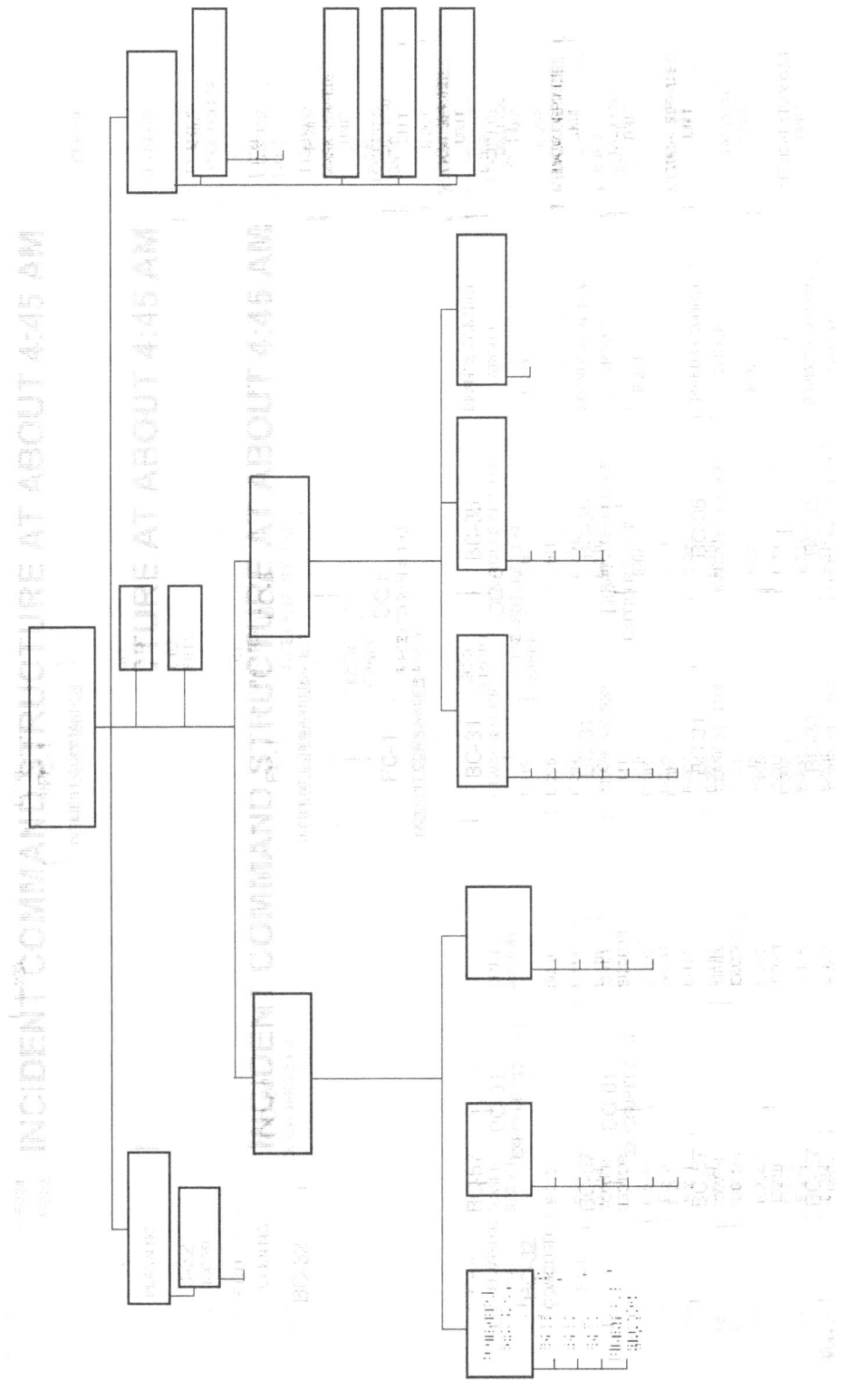

Figure 9. Incident Command Structure at About 4:45 a.m.

Figure 10

Figure 9

Appendix A (continued)

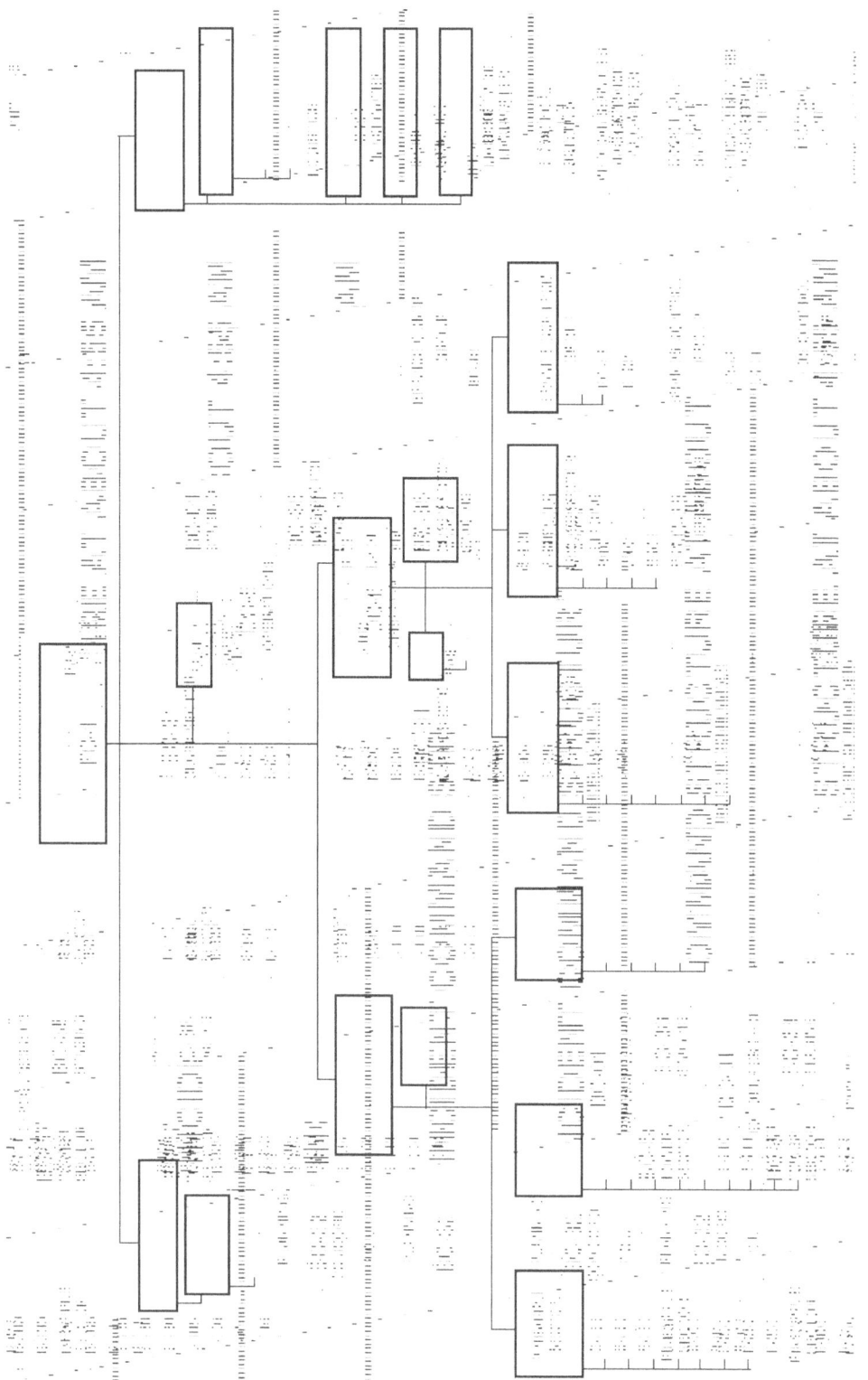

Figure 10. Incident Command Structure at About 4:55 a.m.

Appendix A (continued)

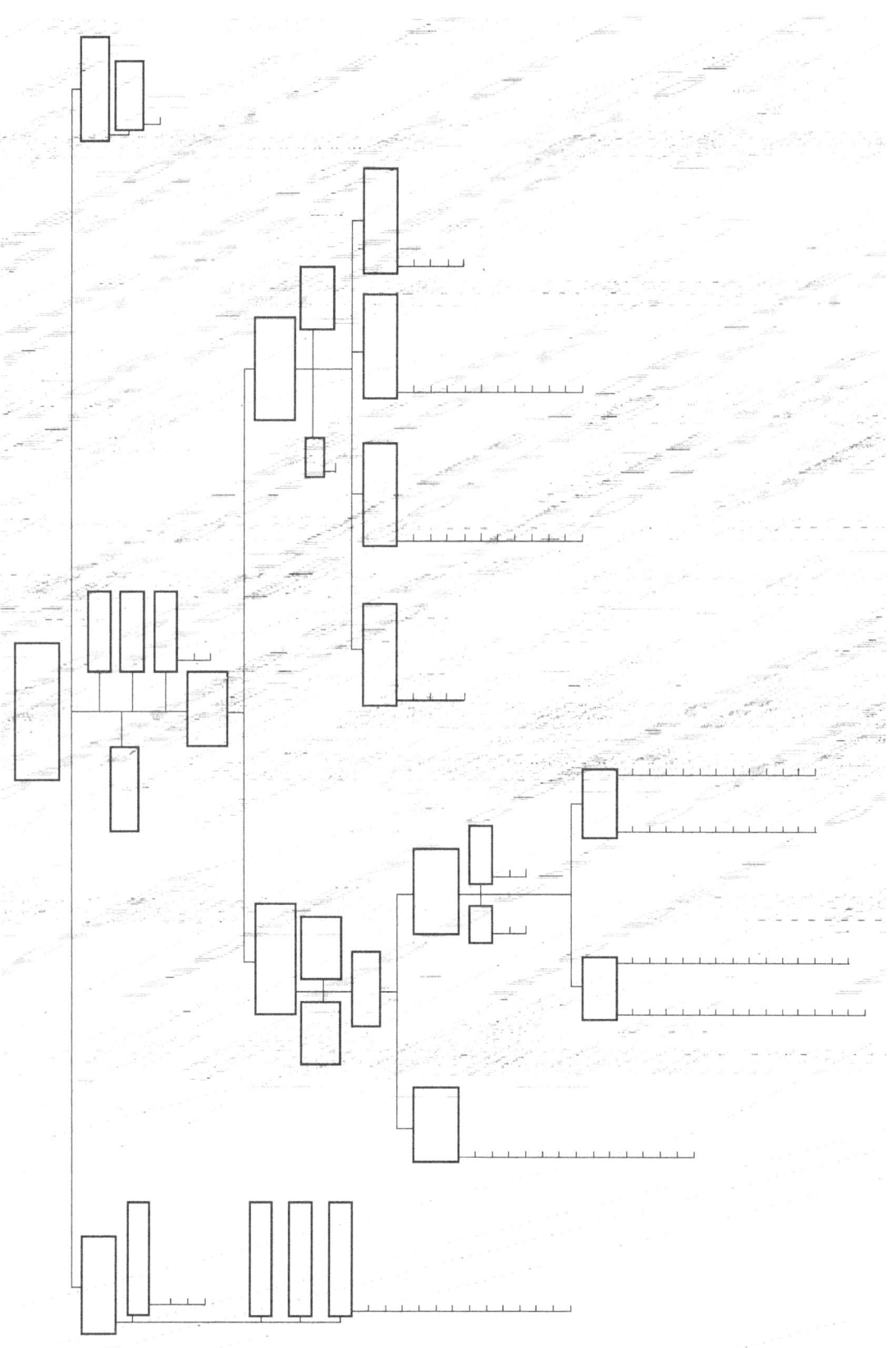

Figure 11. Incident Command Structure for the History of the Fire.
Fire Under Control at 7:09 a.m.

APPENDIX B
Photographs

Except for photograph number 9, all of the photographs were captured from video tape of the fire and fire scene. The video tape was supplied by Eddie McDonald. Photograph number 9 was supplied by F.D.N.Y. Forensic Unit and was taken by T.A. Strandberg. The video captured photographs are available in color from the USFA's Web page as part of the downloadable file.

Appendix B (continued)

Photo 1. Early conditions on the Clark Street (front) side of the Clark Building with Truck 119 in position to operate its elevated master stream.

Photo 2. Later conditions on the front of the Clark Building from street level.

Appendix B (continued)

Photo 3. Conditions from the street

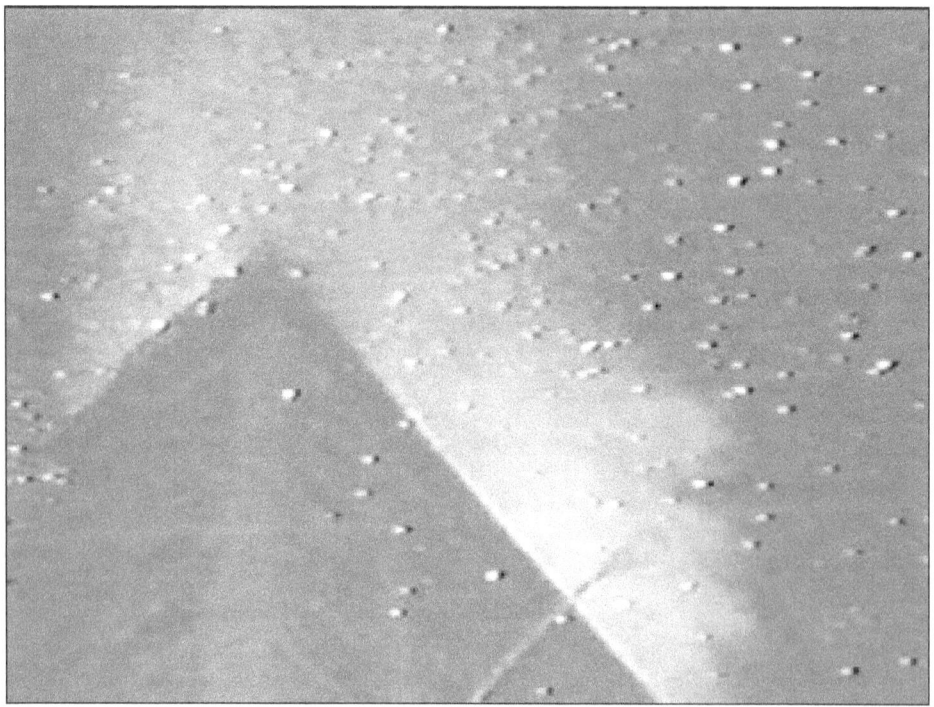

Photo 4. Illustrations of the fire brand conditions over the tops of the buildings.

Appendix B (continued)

Photo 5. The fire, brand, and energy plume from the now totally involved and collapsed Clark Building silhouetted against the rest of the 31 story Tower Building.

Photo 6. Fire conditions at the center of the block behind the Clark Building. The Tower Building is in the background.

Appendix B (continued)

Photo 7. The rear of the Clark Building after the fire. The Tower Building and Grill Building is out of the picture to the right.

Photo 8. The Tower Building and Grill Building with the 31 story Tower Building in the background.

Appendix B (continued)

Photo 9. The Clark Building during demolition as viewed from the rear to the front.

Photo 10. Damage to exposures across Clark Street from the fire building.